German Artillery

1914-1918

by

DAVID NASH

ALMARK PUBLISHING CO., LONDON

First Published—September 1970

ISBN 85524 002 4 (hard cover edition)
ISBN 85524 003 2 (paper covered edition)

Printed in Great Britain by
Vale Press Ltd., Mitcham, Surrey,
for the publishers, Almark Publishing Co.
104-106 Watling Avenue, Edgware,
Middlesex, England.

INTRODUCTION

THIS concise coverage of an extensive subject is intended to explain the organization and equipment of the German artillery of the Great War.

The service was divided into two distinct branches, the Field Artillery, which manned the field guns and light howitzers of the divisional artillery, and the Foot Artillery which served the heavier pieces. Artillerymen were also responsible for the supply of ammunition in the field during the first half of the war. The conditions of actual warfare brought into existence new types of specialist units which are dealt with in the following pages.

Trench mortar units and flak troops are omitted from this volume since they were not under the command of either of the Inspector Generals of Artillery and were not strictly speaking artillery arms. For assistance with the selection of illustrations, acknowledgement is made to the Imperial War Museum. Many of the pictures are taken from German artillery handbooks of the 1914-18 period now in the collection of the Museum.

All the principal weapons used are illustrated though it is not possible to show *every* gun in this present volume. However, though the summary on page 47 lists some 70 different pieces, many of these were obscure or little used types. The relatively small number of guns in widespread service are fully covered here pictorially. Examples of uniforms, camouflage, and typical regimental colours are included in the colour plates.

CONTENTS

A combination of rain, mud, and shell-ravaged ground made the going hard for the artillery gun teams of each side in the 1914-18 conflict on the Western Front. ABOVE: All hands on the drag ropes to negotiate a gap shot into the barbed wire entanglements during the German March Offensive of 1918. Leading driver is wearing the 1915 pattern bluse while the second driver wears an earlier pattern tunic. Team are hauling a C 96 n/A 7.7cm field gun, not visible in the picture (IWM-Q54404). OPPOSITE PAGE: Similar scene on a very muddy track, showing brushwood bundles carried on the gun to provide a 'mat' if the team got too badly bogged down. Note that the leading pair of animals here are oxen, indicative of the severe shortage of horses in the German Army in 1918 (IWM-Q54401).

COVER, TOP: One of the most common camouflage styles applied to German guns was the 'broken colour' pattern and the typical form it took in the 1917-18 period—varying slightly from gun to gun—is shown on this Feldkanone C 96 n/A, the most widely used German artillery piece. COVER, BOTTOM: The moment of full recoil graphically captured by a German official cameraman as a battery of Feldkanone C 96 n/A guns opens fire on British positions during the March Offensive of 1918. Gun crew are positioned exactly as per the drill book, and the door of the limber can be seen on the far side of the gun, positioned as in the diagram on page 10 (IWM-Q23793).

CHAPTER 1

ORGANIZATION OF THE FIELD ARTILLERY

IN 1913 the Field Artillery had an establishment of 642 batteries. These included the field, horse and instructional units composed of field guns and light howitzers which, for all except tactical purposes, were commanded by the Inspector of Field Artillery. At the outbreak of war in 1914, the arm was organized into brigades, each commanded by a *Generalmajor.* These brigades were permanently attached to infantry divisions, whose number they took as a designation, and consisted of two regiments, each of two battalions, each composed of three batteries and a light ammunition column. The batteries were subdivided into three sections each of two guns. The regular regiments were numbered in sequence; the battalions were designated I and II and the batteries, 1 to 3 and 4 to 6.

The Horse Artillery battalions were allotted one to each cavalry division. They consisted of three four-gun batteries and were drawn from regular Field Artillery regiments. One battalion was provided by each of the following units: Guard Field Artillery Regiment No 1, Field Artillery Regiments Nos 1, 3, 5, 8, 10, 11, 12, 15, 35, and the Bavarian Field Artillery Regiment No 5.

The main weapon of the Field Artillery was a field gun, but certain

regiments contained 2nd battalions of light field howitzers. They were arranged so that each Army Corps contained two such battalions.

In addition to the normal field artillery, there was a complete instruction regiment of two battalions attached to the Field Artillery School at Jüterbog, the 'Lehr' Regiment.

The regular artillery regiments in existence in 1914 are listed below:

FIELD ARTILLERY REGIMENTS, 1914

Title	Home station
1. Garde-Feldartillerie-Regiment	Berlin
2. Garde-Feldartillerie-Regiment	Potsdam
3. Garde-Feldartillerie-Regiment	Berlin, Beeskow
4. Garde-Feldartillerie-Regiment	Potsdam
Feldartillerie-Regiment Prinz August von Preußen (1. Litthauisches) Nr. 1	Gumbinnen, Insterburg
1. Pommersches Feldartillerie-Regiment Nr. 2	Kolberg, Belgard
Feldartillerie-Regiment General-Feldzeugmeister (1. Brandenburgisches) Nr. 3	Brandenburg a. H
Feldartillerie-Regiment Prinz-Regent Luitpold von Bayern (Magdeburgisches) Nr. 4	Magdeburg
Feldartillerie-Regiment von Podbielski (1. Niederschlesisches) Nr. 5	Sprottau, Sagan
Feldartillerie-Regiment von Peucker (1. Schlesisches) Nr. 6	Breslau

A 7.7cm C 96 n/A in position to fire down a road. This almost casual placing of guns was typical of the fast-moving March 1918 Offensive when no time was available to erect elaborate pits. Note the wicker ammunition containers.

LEFT: A 7.7cm Feldkanone C 96 n/A, captured by the British in the Somme Offensive, July 1916. It is painted field grey overall. Compare with official drawings on page 12.

Title	Home station
1. Westfälisches Feldartillerie-Regiment Nr. 7	Wesel, Düsseldorf
Feldartillerie-Regiment von Holtzendorff (1. Rheinisches) Nr. 8	Saarlouis, Saarbrücken
Feldartillerie-Regiment General-Feldmarschall Graf Waldersee (Schleswigsches) Nr. 9	Itzehoe
Feldartillerie-Regiment von Scharnhorst (1. Hannoversches) Nr. 10	Hannover
1. Kurhessisches Feldartillerie-Regiment Nr. 11	Kassel, Fritzlar
1. Sächsisches Feldartillerie-Regiment Nr. 12	Dresden
Feldartillerie-Regiment König Karl (1. Württembergisches) Nr. 13	Ulm a. D., Cannstatt.
Feldartillerie-Regiment Großherzog (1. Badisches) Nr.14	Karlsruhe
1. Ober-Elsässisches Feldartillerie-Regiment Nr. 15	Saarburg, Mörchingen
1. Ostpreußisches Feldartillerie-Regiment Nr. 16	Königsberg
2. Pommersches Feldartillerie-Regiment Nr. 17	Bromberg
Feldartillerie-Regiment General-Feldzeugmeister (2. Brandenburgisches) Nr. 18	Frankfurt a. O
1. Thüringisches Feldartillerie-Regiment Nr. 19	Erfurt
1. Posensches Feldartillerie-Regiment Nr. 20	Posen
Feldartillerie-Regiment von Clausewitz (1. Oberschlesisches) Nr. 21	Neiße, Grottkau
2. Westfälisches Feldartillerie-Regiment Nr. 22	Münster
2. Rheinisches Feldartillerie-Regiment Nr. 23	Koblenz
Holsteinsches Feldartillerie-Regiment Nr. 24	Güstrow
Großherzogliches Artilleriekorps, 1. Großherzoglich Hessisches Feldartillerie-Regiment Nr. 25	Darmstadt
2. Hannoversches Feldartillerie-Regiment Nr. 26	Verden

Title	Home station
1. Nassauisches Feldartillerie-Regiment Nr. 27 Oranien	Mainz, Wiesbaden
2. Sächisches Feldartillerie Regiment Nr. 28	Bautzen
2. Württembergisches Feldartillerie-Regiment Nr. 29 Prinz-Regent Luitpold von Bayern	Ludwigsburg
2. Badisches Feldartillerie-Regiment Nr. 30	Rastatt
1. Unter-Elsässisches Feldartillerie-Regiment Nr. 31	Hagenau
3. Sächisches Feldartillerie Regiment Nr. 32	Riesa
1. Lothringisches Feldartillerie-Regiment Nr. 33	Metz
2. Lothringisches Feldartillerie-Regiment Nr. 34	Metz
1. Westpreußisches Feldartillerie-Regiment Nr. 35	Deutsch-Eylau
2. Westpreußisches Feldartillerie-Regiment Nr. 36	Danzig
2. Litthauisches Feldartillerie-Regiment Nr. 37	Insterburg
Vorpommersches Feldartillerie-Regiment Nr. 38	Stettin
Kurmärkisches Feldartillerie-Regiment Nr. 39	Perleberg
Altmärkisches Feldartillerie-Regiment Nr. 40	Burg
2. Niederschlesisches Feldartillerie-Regiment Nr. 41	Glogau
2. Schlesisches Feldartillerie-Regiment Nr. 42	Schweidnitz
Clevesches Feldartillerie-Regiment Nr. 43	Wesel
Triersches Feldartillerie-Regiment Nr. 44	Trier
Lauenburgisches Feldartillerie-Regiment Nr. 45	Altona, Rendsburg
Niedersächsisches Feldartillerie-Regiment Nr. 46	Wolfenbüttel, Celle.
2. Kurhessisches Feldartillerie-Regiment Nr. 47	Fulda
4. Sächisches Feldartillerie Regiment Nr. 48	Dresden
3. Württembergisches Feldartillerie-Regiment Nr. 49	Ulm a. D.
3. Badisches Feldartillerie-Regiment Nr. 50	Karlsruhe
2. Ober-Elsässisches Feldartillerie-Regiment Nr. 51	Straßburg
3. Ostpreußisches Feldartillerie-Regiment Nr. 52	Königsberg
Hinterpommersches Feldartillerie-Regiment Nr. 53	Bromberg, Hohensalza
Neumärkisches Feldartillerie-Regiment Nr. 54	Küstrin, Landsberg
2. Thüringisches Feldartillerie-Regiment Nr. 55	Naumburg
2. Posensches Feldartillerie-Regiment Nr. 56	Lissa
2. Oberschlesisches Feldartillerie-Regiment Nr. 57	Neustadt
Mindensches Feldartillerie-Regiment Nr. 58	Minden
Bergisches Feldartillerie-Regiment Nr. 59	Köln

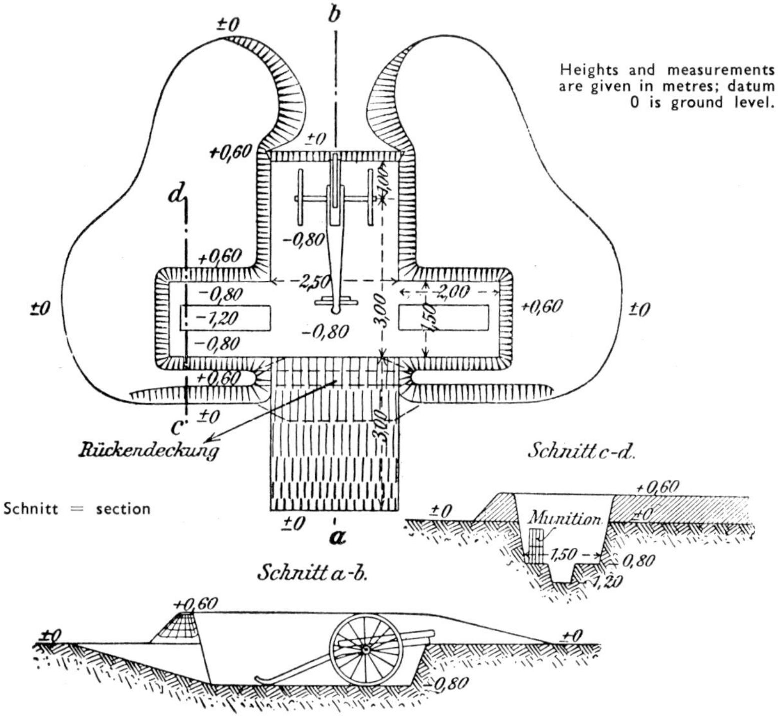

The officially laid down scheme for an earthwork barbette to be constructed when field guns or howitzers formed part of a prepared defensive position. In practice these gun positions varied slightly from this depending on site. Gun entered and left the pit via the rear ramp which was covered by a false deck section while the weapon was emplaced.

Title	**Home station**
Großherzoglich Mecklenburgisches Feldartillerie-Regiment Nr. 60	Schwerin
2. Großherzoglich Hessisches Feldartillerie-Regiment Nr. 61	Darmstadt, Babenhausen
Ostfriesisches Feldartillerie-Regiment Nr. 62	Oldenburg, Osnabrück
2. Nassauisches Feldartillerie-Regiment Nr. 63	Frankfurt Mainz
5. Sächisches Feldartillerie-Regiment Nr. 64	Pirna
4. Württembergisches Feldartillerie-Regiment Nr. 65	Ludwigsburg
4. Badisches Feldartillerie-Regiment Nr. 66	Lahr, Neubreisach
2. Unter-Elsässisches Feldartillerie-Regiment Nr. 67	Hagenau, Bischweiler
6. Sächisches Feldartillerie-Regiment Nr. 68	Riesa
3. Lothringisches Feldartillerie-Regiment Nr. 69	St. Avold
4. Lothringisches Feldartillerie-Regiment Nr. 70	Metz
Feldartillerie-Regiment Nr. 71 Groß-Komtur	Graudenz

Title	Home station
Feldartillerie-Regiment Nr. 72 Hochmeister	Marienwerder Preuß.-Stargard
1. Masurisches Feldartillerie-Regiment Nr. 73	Allenstein
Torgauer Feldartillerie-Regiment Nr. 74	Torgau, Wittenberg
Mansfelder Feldartillerie-Regiment Nr. 75	Halle a. S.
5. Badisches Feldartillerie-Regiment Nr. 76	Freiburg i. Br.
7. Sächisches Feldartillerie-Regiment Nr. 77	Leipzig
8. Sächisches Feldartillerie-Regiment Nr. 78	Wurzen
3. Ostpreußisches Feldartillerie-Regiment Nr. 79	Osterode
3. Ober-Elsässisches Feldartillerie-Regiment Nr. 80	Kolmar
Thorner Feldartillerie-Regiment Nr. 81	Thorn
2. Masurisches Feldartillerie-Regiment Nr. 82	Rastenburg, Lötzen
3. Rheinisches Feldartillerie-Regiment Nr. 83	Bonn, Düren
Straßburger Feldartillerie-Regiment Nr. 84	Straßburg

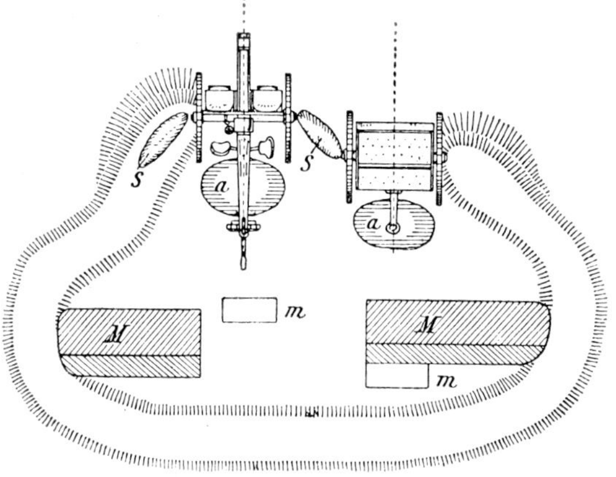

S: sandbags

a: depression scooped in earth for trails

m: ready-use ammunition

M: positions for gun crew (left) and ammunition numbers (right) when not closed up

Standard positions for gun and limber in semi-prepared field conditions. A low earthwork is shown here but the same positions were adopted on open ground—see cover picture. The depressions for the gun and limber trails will be noted and can also just be seen (for the gun) in the cover picture.

BAVARIAN REGIMENTS

Title	Home station
1. Feldartillerie-Regiment Prinz-Regent Luitpold	München
2. Feldartillerie-Regiment Horn	Würzburg
3. Feldartillerie-Regiment Prinz Leopold	Grafenwöhr
4. Feldartillerie-Regiment König	Augsburg
5. Feldartillerie-Regiment König Alfons XIII von Spanien	Landau
6. Feldartillerie-Regiment Prinz Ferdinand von Bourbon, Herzog von Calabrien	Fürth

Title	Home station
7. Feldartillerie-Regiment Prinz-Regent Luitpold	München
8. Feldartillerie-Regiment	Nürnberg
9. Feldartillerie-Regiment	Landsberg
10. Feldartillerie-Regiment	Erlangen
11. Feldartillerie-Regiment	Würzburg
12. Feldartillerie-Regiment	Landau

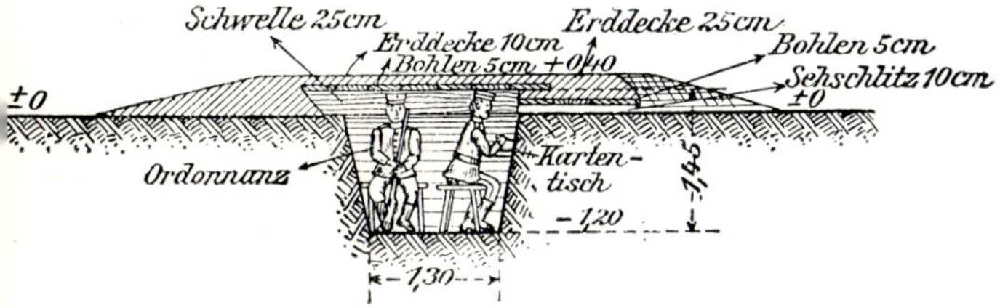

LEFT: Section through battery observation post built into a defensive trench system, as on the Western Front. Entrance was via a tunnel or traverse to a rear corner. *See page 42 for plan.*

In 1915, all batteries were reduced to a four gun standard, thus paving the way for a large expansion in late 1915 and early 1916.

In 1916, the Field Artillery were given new establishments. Owing to the shortage of divisional artillery for the wartime formations of infantry, the strength of artillery units was cut so that more could exist. The regular regiments continued to be brigaded in pairs, but the wartime regiments contained three battalions and were allotted individually to infantry divisions. This gave a varying strength for divisional artillery of 9 or 12 batteries. During this re-organization, a series of independent field batteries numbered 801 to 915 were formed and were used to re-inforce divisional artillery on the Eastern Front. By the end of 1916 the strength of the Field Artillery had risen to some 2,100 batteries.

At about this time, with the introduction of the FK 16 gun, it became the tendency to arm one battery in each battalion with the new gun, resulting in mixed battalions with one battery of 7.7cm C 96 n/A guns, one of 10.5cm howitzers and one of the new FK 16.

By late 1916, the lead horses were withdrawn from munitions and transport wagons. This move was just one effect of a general shortage of horses throughout the entire army. The problem eventually became so acute, that on occasions oxen and even cows were utilized to supplement the over-burdened teams; the unfortunate consequences became apparent during the March Offensive in 1918 when, frequently, the Field Artillery could not keep pace with the infantry advance.

In 1917, the three-battalion regiment became common throughout the Field Artillery, and the war establishment was again revised. Under this new organization, a battery consisted of 6 officers, 21 NCO's, 64 gunners and 45 drivers. In an attempt to save horses, bicycles were issued to batteries. This re-organization set free numbers of batteries drawn from the divisions with a 12 battery complement of field artillery. These surplus units were regrouped into about 80 independent regiments and formed a General Headquarters Reserve.

By the end of the war, the number of Field Artillery batteries had risen to some 2,900.

CHAPTER 2
FIELD ARTILLERY WEAPONS

THE Field Artillery employed two principal types of field gun, the models of 1896 and 1916, both of 7.7cm calibre.

The *Feldkanone 96 n/A* (or *C 96 n/A*) was first introduced into service in 1896. It was subsequently modified in 1905 and 1906 and was the main weapon of the Field Artillery until the introduction of the FK 16.

The gun weighed 1,930 pounds and was mounted on a shielded recoil carriage. The barrel was rifled with 32 grooves, the rear end was shaped to take the single motion wedge breech, and three recoil

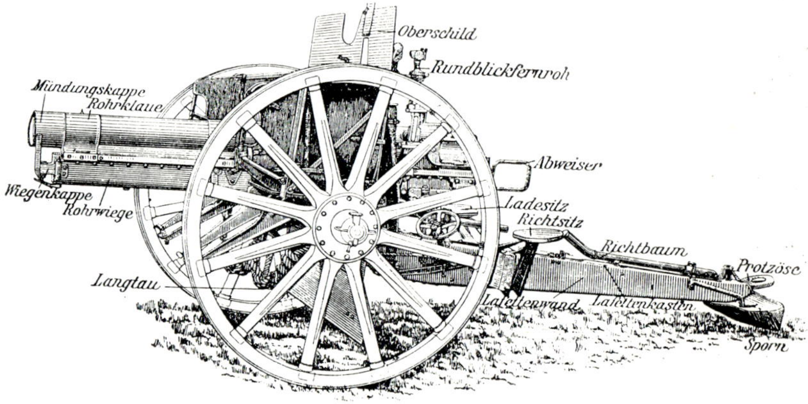

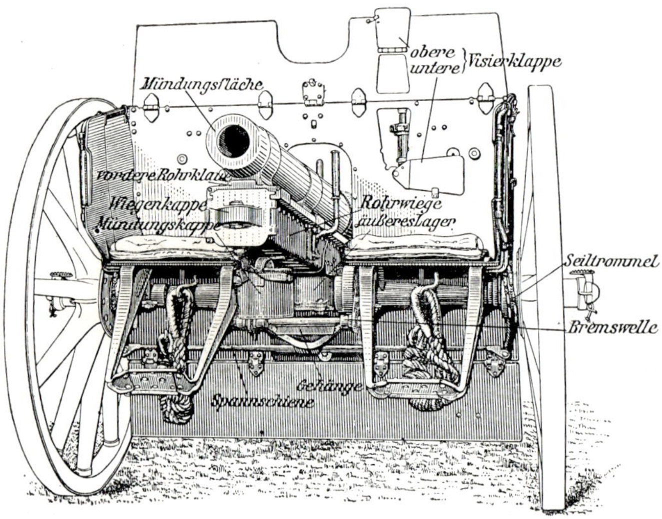

Drawings from the German field artillery manual show detail features of the 7.7cm Feldkanone C 96 n/A, principal weapon of the field artillery batteries. Note drag rope stowage on axle casings.

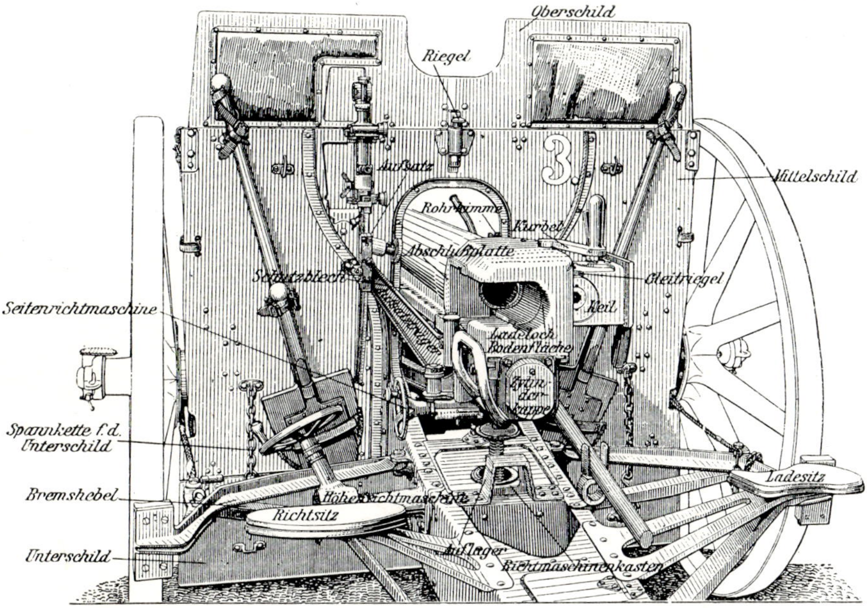

View of the 7.7cm C 96 n/A from the trail end. Note the individual number (3) of the gun within the battery. This was the only identity marking retained during the war years. The hinged upper shield (oberschild) was raised and locked into place when the gun was in action. Note the padded leather sections to protect the heads of layer and breech worker respectively who sat on the trail seats. Stowage of shovels and entrenching tools on shield should also be noted.

guides were shrunk on to the gun, enabling it to recoil along the buffer slide. The carriage consisted of a cradle, housing the buffer and running-out springs, which were directly under the barrel, a box trail fitted with a spade and traversing lever, and a shield which was 5½ft high when fully extended. Two axletree seats enabled the layer and loader to travel with the piece, their backs to the front of the shield. Horse artillery batteries used this same gun but did not utilize the seats. The carriage was mounted on 4ft 5in wheels with a 5ft track.

The gun's maximum range was 9,186 yards, which could be obtained with a percussion-fused shell. When firing time-fused ammunition, the range achieved was approximately two thirds of the maximum. The muzzle velocity with a normal charge firing a *K.Gr. 15* projectile was 1,526ft per second. The limits of traverse were 7°15′ to left or right and elevation was from +16° to −12°.

The limber carried 36 rounds of ordinary shell or 24 long shells plus 6 ordinary. The seat on top of the limber accommodated three gunners.

The *Nahkampf* batteries used the C 96 n/A, mounted on lower wheels, 43in in diameter. Equipped for an anti-tank role, special armour piercing projectiles were provided for use by these batteries.

Towards the end of 1916, a new field gun was taken into service. It was basically a modified C 96 n/A mounted on a field howitzer carriage. Its designation was the *K.i.H. (Kanone in Haubitzlafette).*

A 7.7cm FK 16 battery preparing to open fire on the Somme. This gun replaced the C 96 n/A in the latter part of the war and had a longer barrel plus the carriage of the Feldhaubitz 98/09, shown opposite. Note varied attire of crew with some in boots and others in puttees.

Subsequent modifications of this gun produced the *Feld-Kanone 16* (FK 16), with which many batteries were re-equipped.

The FK 16 was about 2 feet longer than the 1896 model and was mounted on the 98/09 field howitzer carriage, thus making the gun capable of a greatly increased elevation. The breech block was identical to that used on the C 96 n/A. The powder chamber was, however, 80mm longer allowing for a more powerful propellant charge. The upper and lower shields were movable, the latter being designed to be hooked up. Provision was made for the fitting of a circular platform, to be placed under the wheels, which enabled the gun to be manually traversed without limit and also made for greater stability. With an additional charge, a muzzle velocity of 1,968ft per second and a range of 11,264 yards could be obtained. The gun weighed 2,750 pounds and the limits of elevation were +38° to −9°30′, traverse was 2° to left or right. The limber was of the standard design issued to the Field Artillery.

The standard field howitzer was the old 1898 pattern weapon

A 7.7cm FK 16 in action on the Eastern Front. Gun is on full recoil. During barrage fire, as in views on this page, limber was usually withdrawn to rear in favour of large dump of ammunition built-up for sustained fire periods. This supply was maintained by ammunition parties in rear.

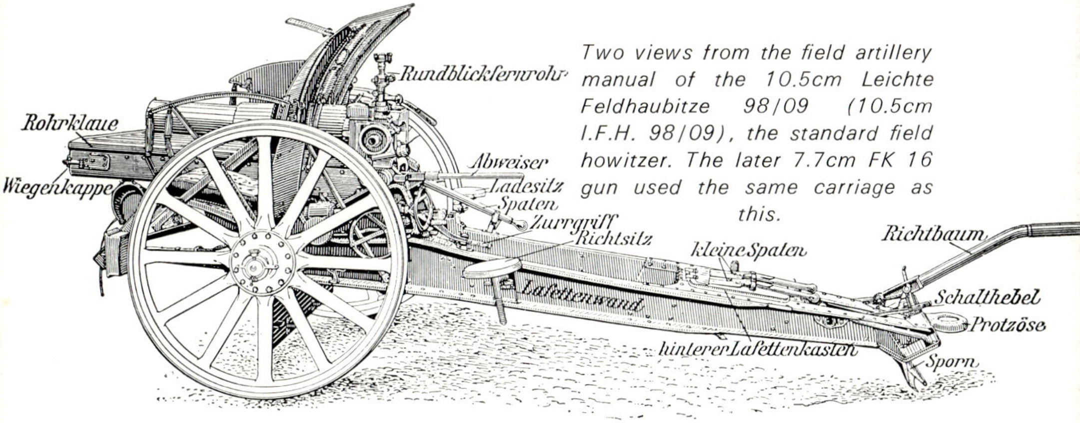

Two views from the field artillery manual of the 10.5cm Leichte Feldhaubitze 98/09 (10.5cm l.F.H. 98/09), the standard field howitzer. The later 7.7cm FK 16 gun used the same carriage as this.

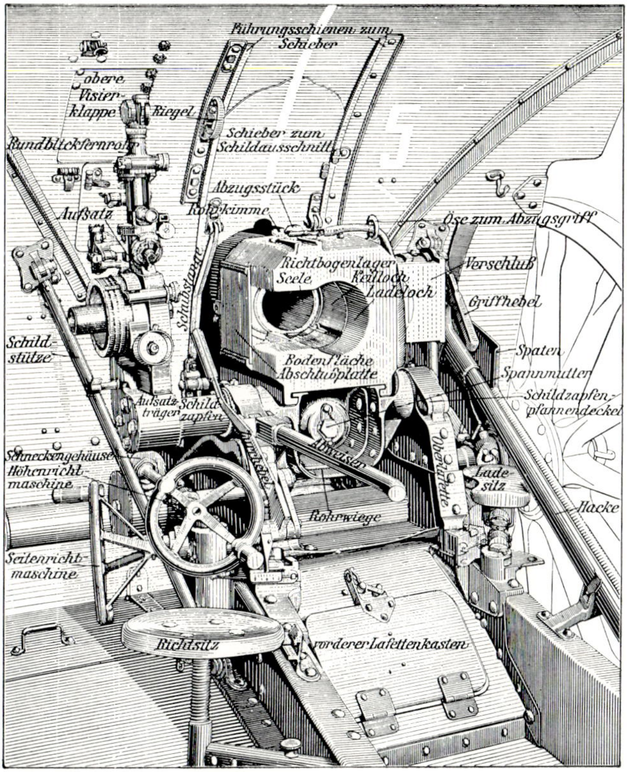

entirely re-modelled and mounted on a shielded recoil carriage. This improved weapon was known as the *leichte 10.5cm Feldhaubitze 98/09*. It had a single motion wedge breech block and the bore was rifled with 32 grooves. The wheels were 4 feet in diameter and had a track of 5 feet. The limits of elevation were −13° to +40° and there was 4° traverse. With a full charge, a muzzle velocity of 991ft per second and a range of 7,655 yds could be obtained. Weight in action was 22½ cwt.

Two newer light field howitzers were introduced during the war,

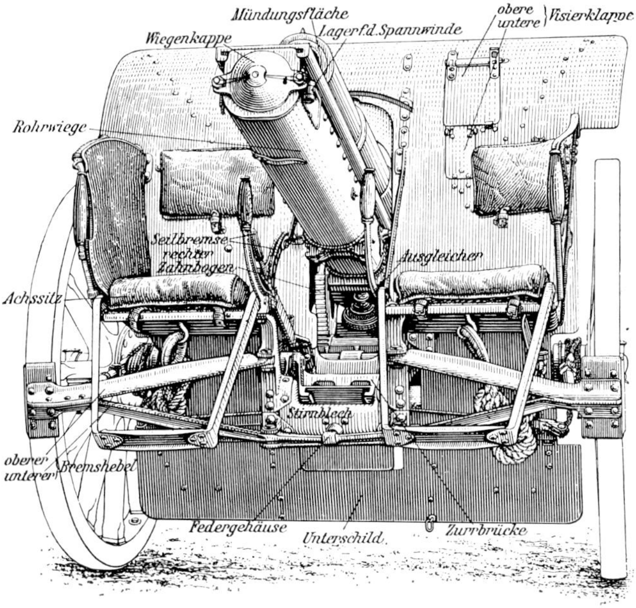

Front view of the 10.5cm Leichte Feldhaubitze 98/09. Note that on this carriage the drag ropes were carried under the riding seats. Brakes were also operated from these seats.

the *l.F.H. 16* and *l.F.H. Kp.* Both were mounted on the 98/09 carriage and both had a longer barrel than the *l.F.H. 98/09.* Maximum ranges of 10,936 and 11,210 yards could be attained by the '16 and Krupp *(Kp.)* models respectively.

The limbers in use with these howitzers each carried 24 rounds of ammunition.

NB: Drawings are not to common scale

TOP: 7.7cm Feldkanone C 96 n/A shown in the peacetime blue-grey colour. From 1915 this shade was gradually replaced by field grey, actually a greenish grey colour. ABOVE: A Leichte Feldhaubitze 98/09 in basic field grey with the disruptive 'blotch' scheme of camouflage consisting of dark green and light grey patches.

OPPOSITE: The limber (feldprotze) for the 7.7cm Feldkanone 96 n/A. Note the wicker containers for the rounds, the riding bench, and the valise. Limber for the Leichte Feldhaubitze 98/09 was almost identical except that the rounds were stowed vertically and the box was slightly deeper.

RIGHT: The gun limber (Feldprotze) used with the 7.7cm C 96 n/A field gun. Note the folding support to keep it horizontal for loading up. LEFT: The standard ammunition wagon (Munitionshinterwagen) of the field artillery of which four were generally allocated to each battery. They were also used by ammunition columns.

WAR ESTABLISHMENT OF FIELD ARTILLERY UNITS, 1914

	Officers and Officials	Other Ranks	Horses	Guns, 6-Horsed	Bty. Ammo. Wagons 6-Horsed	Observation Wagon 6-Horsed	Other Vehicles
HORSE BATTERIES	4	120	170	4	4	1	4
FIELD BATTERIES	5	156	136	6	6	1	4
LIGHT MUNIT. COLUMN	4	188	178	–	–	–	24
HORSE BATTALION	24	574	719	12	12	4	37
FIELD A. BATTALION	27	681	621	18	18	4	37
FIELD A. REGT.	59	1,371	1,258	36	36	8	75
FIELD A. BRIGADE	120	2,760	2,528	72	72	16	151

WAR ESTABLISHMENT OF A FIELD ARTILLERY BATTERY, 18. 4. 1916

PERSONNEL			HORSES			VEHICLES		
				Draught				
Officers	Other Ranks		Riding	Wheel	Lead	6-Horse	4-Horse	2-Horse
1	–	BATTERY COMMANDER	2					
3	–	LIEUTENANTS or 2nd LTS.	6					
	1	BATTERY SGT. MJR.	1					
	1	DEPUTY BSM	1					
	1	ENSIGN	1					
	8	CORPORALS	8					
	3	TRUMPETERS	3					
	12	BOMBARDIERS	4					
	2	FITTERS						
	2	SADDLERS						
	1	WHEELER						
	33	GUNNERS						
	36	DRIVERS						
	9	TRAIN DRIVERS						
	1	SHOEING-SMITH	1					
	1	MEDICAL CORPORAL						
	1	ARTIFICER						
		4 GUNS or HOWITZERS with LIMBERS		8	16	4		
		4 AMMUNITION WAGONS } 1 OBSERVATION WAGON }		10	20	5		
		2 STORES WAGONS		4	4		2	
		1 FORAGE WAGON		2	2		1	
		1 SUPPLY WAGON		2			1	
		1 FIELD KITCHEN		2				1
		SPARE HORSES		2	2			
4	112	**TOTAL**	27	30	44	9	4	1

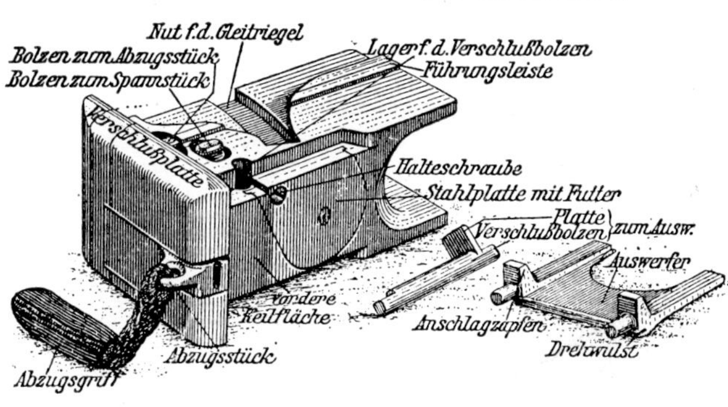

RIGHT: A close view of the sliding breech mechanism of the Feldkanone C 96 n/A reproduced from the official handbook.

Colour of 1st West Prussian Foot Artillery Regiment No 11

Colour of 14th Baden Foot Artillery Regiment

Colour of the Garde Fuss-artillerie Regiment

NB: Streamers on colours normally indicated battle honours and awards won by regiment

Colour of Fuss-artillerie Regiment Von Singer (Ostpreuss) Nr. 1

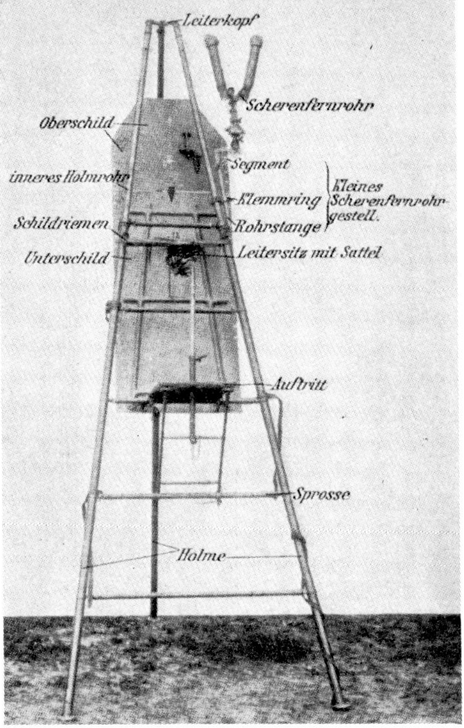

The Beobachtungswagen für feldartillerie (observation wagon) was supplied one to each battery. These views show the observation equipment (Beobachtungsleiter) stowed in the wagon (BELOW), erected on the ground (standing height) (UPPER LEFT), and erected on the wagon (RIGHT).

In addition to the positions shown here, the observation equipment could be used on the ground from the prone position. The shield was turned on its side and the sight was fitted to a low tripod mount.

Two alternative positions for the observation equipment and shields are shown here, allowing the observation officer a flexible choice of spotting height depending on location. Field telephone equipment was carried and the line was paid out to the rear over the sheave and arm clearly shown on the opposite page.

Colour of 12th Saxon Field Artillery Regiment

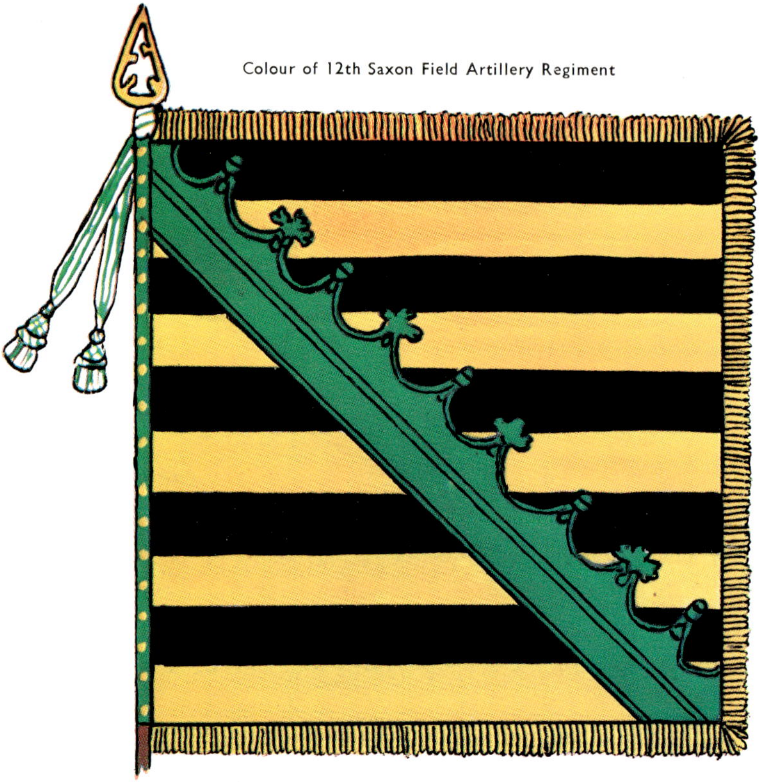

NB: Plate shows front and back views of same colour

19th Regiment had similar colour but with number 19 inside the grenade symbols

Artillery Uniforms, 1914-16

LEFT to RIGHT: Foot Artillery private in forage cap and typical order as member of gun crew. Foot Artillery private in pickelhaube with drab helmet cover and in full marching order. The Brandenburg cuffs indicate Foot Artillery. Private and Unteroffitzier of Field Artillery. Note the Swedish style cuff, which distinguished the Field Artillery Regiments.

Artillery Shoulder Straps, 1916-18

1 2 3 4 5

By order of the Kaiser in September 1915, coloured shoulder straps were introduced to replace those of a field grey colour hitherto worn (and illustrated on page 28). These new straps were only gradually issued but by 1917 were in common use. Key: (1) Field Artillery. (2) Foot Artillery. (3) Shallmesstruppen—sound-ranging troops. (4) Artilleriemesstruppen—observation post troops. (5) Field Artillery School. Note that there were variations in pattern depending on the manufacturer.

The 15cm s.F.H. (Schwere Feld Haubitze) '02 being emplaced for firing in 1918. Note the 'blotch' style of camouflage, and the absence of a shield on this model. Varied uniform styles show jackboots, puttees, or neither, and is typical of the period.

CHAPTER 3

ORGANIZATION OF THE FOOT ARTILLERY

ALL Foot Artillery units were subordinated to the Inspector General of Foot Artillery. The Inspectorate was divided into three inspections each controlling two brigades each of three or four regiments. One regiment was attached to each Army Corps district.

In 1913, the total Foot Artillery strength was 48 battalions, of which 38 were drawn from Prussia, 6 from Bavaria and 4 from Saxony. The 24 regiments were normally organized from two battalions, each of 4 batteries, excepting those battalions composed of 21cm Mortars, which had only 2. Each battery had a complement of 4 guns. In addition to the field units, there was a Foot Artillery School situated at Jüterbog, with its permanently attached 'Lehr' regiment of two battalions.

The normal peace time allotment of Foot Artillery was one battalion of 15cm Howitzers or 21cm mortars to each army corps, the remainder being attached as army troops. Upon mobilization, however, the quantity of heavy artillery was augmented by the formation of siege artillery units. These were organized in regiments of two battalions, each of four batteries containing either 24 guns or 16 howitzers. The siege regiments were armed with a variety of pieces not to be found in the pre-war establishments. The heavy guns were the 10cm gun '04; heavy 12cm gun; 13cm gun; long 15cm gun; heavy field howitzer '02, and the 21cm bronze mortar.

The mobilization of the various classes of reservists in 1914 activated first the Reserve Regiments and then the Landwehr Regiments, similar to the process carried through in the Field Artillery. The very nature of the Great War, however, dictated that eventually the rate of expansion of the Foot Artillery would outstrip the lighter field artillery and, in some cases, *Landsturm* and *Ersatz* battalions, all drawn from the regimental depots, saw service in the front line; also, the number of batteries within a battalion in some cases rose to five.

In 1915, many independent batteries were organized numbered from 101-150 and 200-800. A number of these units were equipped with captured or obsolete guns and they were intended for a static role. At the same time, independent battalions numbered from 21 upwards were formed and a year later a second series of these battalions numbered upwards from 40 were added to the growing Foot Artillery. These battalions were of three of four batteries and a number of them were formed merely by taking units from the 200-800 series and giving them horses and a new designation.

Wartime battery organization was dependent, to a very large extent, on the calibre of the guns. A 21cm battery in 1918 consisted of three guns, six officers, 200 other ranks, 100 heavy and 25 light draught horses. At the same time a 15cm Howitzer battery had 4 guns, 4 officers, 120 other ranks, but not all batteries of this class even had horses.

At the beginning of 1918, the Foot Artillery contained some 2,250 batteries, an increase of five or six times over the pre-war strength.

A field howitzer battery on the move to take up new positions at Kemmel in 1918. Note the typical disruptive camouflage. Weapon shown is the lg. 15cm s.F.H. '13. Note heavy trail, folding shield, and wide tyres.

Artillery Uniforms

LEFT: Leutnant and Private in typical uniforms of 1917-18, both wearing the 1915 pattern Bluse which had concealed buttons and plain turned up cuffs. The private, from the Foot Artillery, has the 1915 pattern shoulder straps, a camouflaged 1916 pattern steel helmet, and wears puttees instead of jackboots. Note the anti-gas respirator carried in a cylindrical container. RIGHT: A private of Field Artillery in the full dress parade uniform of 1914.

FOOT ARTILLERY REGIMENTS, 1914

Title	Home station
Garde-Fußartillerie-Regiment	Spandau
Fußartillerie-Regiment von Linger (Ostpreußisches) Nr. 1	Königsberg, Feste Boyen-Lötzen
Fußartillerie-Regiment von Hindersin (1 Pommersches) Nr. 2	Swinemünde, Emden, Insel Borkum
Fußartillerie-Regiment General-Feldzeugmeister (Brandenburgisches) Nr. 3	Mainz
Fußartillerie-Regiment Encke (Magdeburgisches) Nr. 4	Magdeburg
Niederschles. Fußartillerie-Regiment Nr. 5	Posen
Fußartillerie-Regiment von Dieskau (Schlesisches) Nr. 6	Neiße, Glogau
Westfälisches Fußartillerie-Regiment Nr. 7	Köln
Rheinisches Fußartillerie-Regiment Nr. 8	Metz
Schleswig-Holsteinisches Fußartillerie-Regiment Nr. 9	Ehrenbreitstein, Köln
Niedersächsisches Fußartillerie-Regiment Nr. 10	Straßburg
Westpreußisches Fußartillerie-Regiment Nr. 11	Thorn

1. Sächisches Fußartillerie-Regiment Nr. 12	Metz
Hohenzollernsches Fußartillerie-Regiment Nr. 13	Ulm, Neubreisach
Badisches Fußartillerie-Regiment Nr. 14	Straßburg
2. Pommersches Fußartillerie-Regiment Nr. 15	Bromberg, Graudenz
Lothringisches Fußartillerie-Regiment Nr. 16	Diedenhofen, Müllheim/ Baden
2. Westpreußisches Fußartillerie-Regiment Nr. 17	Danzig-Neufahrwasser, Pillau
Thüringisches Fußartillerie-Regiment Nr. 18	Mainz
2. Sächisches Fußartillerie-Regiment Nr. 19	Dresden
Lauenburgisches Fußartillerie-Regiment Nr. 20	Itzehoe

BAVARIAN REGIMENTS

Title	**Home station**
1. Fußartillerie-Regiment von Bothmer	München
2. Fußartillerie-Regiment	Metz
3. Fußartillerie-Regiment	Ingolstadt

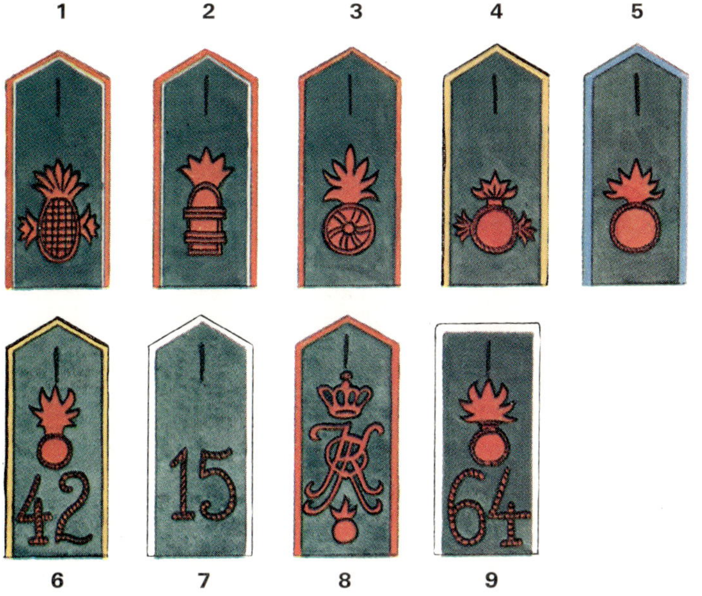

Patterns of shoulder straps worn in 1914 with the 1910 Field Service uniform. Figs 1, 3, 4, 5 are those of Guard Field Artillery Regts 1-4 respectively. Fig 2 is the shoulder strap of the Horse Artillery detachment of Guard Field Artillery Regt No. 1. Fig 6 is from Field Artillery Regt No. 42. Fig 7 is of Foot Artillery Regt No. 15. Fig 8 is an example of a regimental cypher, in this case of Field Artillery Regt No. 13. Fig 9 shows a Saxon shoulder strap, distinguished by the square cut top; it is of Field Artillery Regt No. 64.

CHAPTER 4
FOOT ARTILLERY WEAPONS

THE armament of the Foot Artillery was composed approximately of 25% guns and 75% howitzers. The most common types of howitzers in use were of 15cm calibre and these constituted some 50% of the total armament of the Foot Artillery.

There were four patterns of the *15cm Schwere Feld Haubitze,* the *s.F.H.;* the *s.F.H. '02;* the *s.F.H. '13* and the *lg. s.F.H. '13.*

The *s.F.H.* was mounted on an old type of heavy limbered carriage and recoil was checked by means of tyre brakes. This howitzer was not necessarily fired from a platform, but soft ground mats were frequently placed under the wheels and trail. The limit of elevation was +65° and muzzle velocity on the full seven charges was 905ft per second. Mamimum range was 6,616 yards. The barrel was 10.8 calibres long.

The *s.F.H. '02* was mounted on a recoil carriage with spade, buffer and running-out springs. The barrel was 12 calibres long. The carriage was not shielded. The limit of elevation was +42° and lateral traverse was 3°56′. Muzzle velocity on the full six charges was 1,066ft per second and the range was 8,147 yards.

The *s.F.H. '13* was an improved and shielded version of the '02 model with a barrel 14 calibres long. The limit of elevation was +45°, muzzle velocity was 1,196ft per second and the maximum range was 9,296 yards.

The *lg. 15cm s.F.H. '13* had a barrel 17 calibres long over which was shrunk a muzzle ring, two guide rings and a breech jacket. The breech was of the sliding wedge type. The trunnions were vertically above the axle. The trail, which was noteworthy because of its length, was fitted with a hinged shield and a folding spade. The wheels were 49 inch in diameter and were fitted with 4½ inch tyres. The limit of elevation was 45° and the maximum range was 9,296 yards.

Next to the 15cm howitzers the most common of Foot Artillery pieces were the 21cm mortars, forming about 25% of the total Foot Artillery armament. The term *Mörser* was used by the German Army to describe

A 15cm s.F.H. '13 being prepared for action at Kemmel in 1918. On the cradle can be seen a personal inscription crudely applied. Note the heavy recoil spade which was carried on the limber (see picture, page 27) when on the march.

A 15cm field howitzer (s.F.H. '13) in action. This photo shows the barrel on full recoil, graphically demonstrating the purpose of a box trail.

most of its howitzers of 21cm calibre or over. There were three 21cm *Mörser* patterns.

The *21cm Mörser* (or *21cm Mrs.*) dated from 1902, but was still in use in 1918, even though it was considered to be an obsolete piece. It had a carriage provided with travelling wheels. When in action the carriage rested on two layers of baulks of timber. The recoil was checked by two movable inclined planes which were placed on each side of the carriage so that the firing tracks had to run up them when the carriage recoiled. Its maximum range was 8,968 yards and the muzzle velocity was 1,158ft per second. The length of the barrel was 10 calibres.

The *Mörser (Mrs.)* was introduced in 1910 and was the standard weapon of its type throughout the war. It had a crank breech and was

A 21cm Mörser being manhandled into position. This piece is painted in the crude fashion of camouflage mentioned in the text. Note the wood and steel 'shoes' attached to the wheels for crossing muddy terrain.

Officer and NCO of the Garde Feldartillerie Regt No. 1. They are wearing the field grey full dress uniform introduced in Sept 1915 to replace the old blue pattern.

OPPOSITE RIGHT: A close up view of a 21cm Mörser showing the sights and breech mechanism. FAR RIGHT: The 21cm Mörser, 1902 model. Compare with 1910 model, shown on previous page, which had a hooded shield.

BELOW: A battery of lg. 15cm s.F.H. '13 field howitzers limber up to move out of their positions in the Hindenburg Line during the closing months of the war. Note the hinged shield and the ammunition wagon (see page 18) used as a limber.

A 21cm Mörser position. Ammunition is in readiness near to the piece and the wicker cases in which the shells were packed are clearly shown. The crew are engaged in traversing the piece, by means of the handspike, under the direction of the layer who is looking into the sight.

mounted in a cradle and recoiled on its carriage. The two hydraulic buffers and the running-out springs were mounted above the barrel. The gun detachment was protected by a special hooded shield. The muzzle velocity on a full charge was 1,203ft per second and the range was 10,288 yards.

The long mortar, *lg. 21cm Mrs,* was used in small quantities towards the end of the war. As its name implies, the barrel was longer than the two previous models and its range was increased to 11,155 yards.

The remainder of the Foot Artillery's armament consisted of a large number of guns and heavy howitzers, ranging in calibre between 9cm and 42cm. Most common were pieces of 10cm, 13cm and 15cm calibres. 24cm naval guns were used to a limited extent and many other guns of naval origin, either spares, taken from old warships or foreign orders undelivered at the outbreak of war; in addition, Austrian, Belgian, French and Russian pieces were all used in the field. Many of the German-produced weapons dated from before the war and the number of obsolete types in use actually increased, rather than decreased, as the war progressed.

There were four main patterns of 10cm gun, known respectively as

A battery of 10cm 04/14 guns of the Foot Artillery in action. The gun in the foreground is on full recoil. Note the men's rifles and personal equipment piled in the immediate foreground.

10cm K, 10cm K.97 10cm K.04 and *10cm K.04/14.* Three other variants in this calibre also existed, designated according to their mountings. One type had an overhead shield, whilst both of the others—one of which was a re-inforced gun—were mounted in turrets.

The *10cm K.04* was mounted on a recoil carriage fitted with a hydraulic buffer, running-out springs and a trail with spade. There was no shield and the breech was of the single motion vertical wedge type. The limits of elevation were $-5°$ to $+30°$ and lateral traverse was 3°56′. Maximum percussion range was 11,264 yards and the muzzle velocity was 1,830ft per second.

The *10cm K.04/14* made its first appearance in the middle of 1915 and was in general use by 1917. It differed from the '04 model in that its longer barrel measured 35 calibres. Its maximum range was 12,085 yards.

The 13cm gun was a long range high velocity QF gun which was introduced during the war to replace the 15cm long gun. It was mounted on a shielded carriage fitted with a hydraulic recoil system, the buffer and springs being situated in a cradle beneath the barrel. The trail was fitted with a small fixed spade and a second, much larger, hinged one. The wheels were made of either wood or steel. The limit of elevation was $+26°$ and the maximum percussion range was 15,748 yards. Muzzle velocity was 2,280ft per second.

BELOW: A closer view of a 10cm K. 04/14, in action on the Somme late in 1917. Wheels are supported on timber baulks.

A 15cm naval gun mounted on a canal barge. Germans used some guns like this in Belgium taking advantage of the mobility of the canal system.

There were numerous types of 15cm guns in use. The 15cm ring cannon *(15cm R.K.)* had a barrel 23 calibres long. Its limits of elevation were from $-5°$ to $+37°$ and a maximum range of 8,640 yards could be obtained with a percussion fused shell. A long version of this gun was also used, the *lg. 15cm R.K.,* which had a barrel length of 30 calibres. Its range was 7,655 yards.

The long 15cm gun *(lg. 15cm K.)* had a barrel 30 calibres long and a maximum range of 10,936 yards.

The 15cm experimental gun *(15cm Vers. K.i.R.L.)* made its début in 1916 and saw considerable service in the field. Firing a long shell, a range of 21,000 yards could be obtained, but with normal ammunition, only 18,400 yards could be expected. The *15cm K.16* was a development from the above weapon with a range increased to nearly 25,000 yards.

Two 15cm naval guns were in use as long range QF heavy field pieces. The *15cm S.K. L/30* and *L/40,* differing only in barrel length, had ranges of 13,233 and 20,451 yards respectively. They were mounted on wheeled carriages and were drawn by motor traction engines.

The comparatively rare heavy howitzers were of 28cm, 30.5cm and 42cm calibres. Two types of 28cm howitzer were in use; one was mounted on a travelling carriage, the other was a coastal defence weapon.

In the 30.5cm calibre, there were two coastal defence mortars, the *s. Kst. Mrs.* and the *s. Kst. Mrs. 09.* Both mortars had ranges of about 10,000 yards. In addition to these, there were three field variations.

The 30.5cm Austrian 1911 model had a barrel which recoiled within a circular cradle made of cast steel. Above the cradle were two buffer cylinders and below it were three cylinders, housing the hydro-pneumatic recuperator. The cradle was mounted on trunnions on a carriage formed by two uprights braced in front, and fitted to a rectangular steel platform. The whole equipment was carried on three trucks towed by a 100 h.p. Daimler tractor. Its total weight was about 24 tons; the limits of elevation were $+45°$ to $+75°$ and it threw an 838lb shell some 10,500 yds. The rate of fire was one round per six minutes. It was this weapon which vanquished the Liege forts in 1914.

ABOVE: A 17cm K.i.R.L. on a railway carriage. This piece was not a true railway gun, although another version of the same weapon was used in this rôle. FOOT OF PAGE: 17cm K.i.R.L. This was the wheeled version of the 17cm Railway gun and is the same piece as can be seen mounted on a railway carriage in the top picture. Just visible is a traversing channel, embedded in the ground to take the trail.

Krupp produced a similar mortar, some 20 of which saw service with the German Army. In 1914, the few of these weapons that existed, were manned by Krupp's employees and not by the army.

The heaviest mortar in use was of 42cm calibre. It was designated a short naval gun *(Kz. Mar. K.)* and had a range of 18,000 yards.

The *17cm S.K. L/40* was of naval origin and was the largest piece to be mounted on a wheeled carriage. Its range was 29,200 yards. Owing to its size, this piece was frequently carried on, and fired from, a flat railway carriage. Two other versions of the same gun existed, one a true railway gun, the *17cm K. Eis.,* and the other, the *17cm K. Bett-Gesch,* which was fired from a fixed mounting.

All heavier pieces were either on fixed or railway mountings. Two naval railway guns were the *21cm S.K. L/45 Eis.* and the *24cm S.K.*

ABOVE: The 15cm K. L/40 was another naval gun used by the army on a field carriage. For travelling it was split, as shown here, with the barrel transported on a separate carriage. Note the seat for the brakeman on the carriage. Most heavy guns were transported this way to keep the loads to manageable sizes.

WAR ESTABLISHMENTS OF FOOT ARTILLERY UNITS, 1914

	Officers & Officials	Other Ranks	Horses	Guns	Ammunition Wagons	Other Vehicles
HEAVY FIELD HZ BTY.	6	225	122	4	8	7
MORTAR BTY.	6	264	148	4	–	19
HEAVY FIELD HZ BTL.	37	1,182	707	16	56	37
MORTAR BATTALION	25	810	515	8	24	47
REGT'L. STAFF.	4	24	13	–	–	1

GUN STRENGTH OF FOOT ARTILLERY BATTERIES, 1918

Weapon	Quantity
10cm GUN	4
13cm GUN	2
15cm GUN	2
15cm HOWITZER	4
21cm HOWITZER (MORTAR)	3
9cm GUN	6, 8 or 12

L/40 Eis. The latter was also used on a fixed platform, the *24cm S.K. Bett-Gesch.*

The 35.6cm coastal defence gun *(35.6cm S.K. L/52)*, made by Krupp, was set into a concrete base. One such gun was situated south-west of Lille and was fixed to a 10 feet thick slab of concrete measuring 45 x 55 feet on the surface. This particular gun was capable of firing a shell over a distance of 50,300 yards.

The heaviest naval gun employed on a land mounting was the *38cm S.K. L/45.* This gun was mounted in a concrete pit, across half of which was built a steel girder construction to act as a mount. Its range was 46,000 yards.

TOP: Two views of the 38cm S.K. L/45, 'Long Max'. This gun was emplaced and was not a railway piece as is perhaps indicated by the photograph. The rails were used for the transport of its huge shell.

ABOVE: The 38cm shell being taken to 'Long Max' on a narrow gauge railway trolley. Nearest man wears the 1915 pattern Bluse over fatigue overalls.

OPPOSITE PAGE: Two views of the field artillery forage wagon as issued originally to all batteries equipped with the C 96 n/A. Note the locker under the footboard and at the sides. Stores were carried in the front of the wagon ahead of the portable forge.

CHAPTER 5

AMMUNITION COLUMNS AND SUPPORTING EQUIPMENT

THE ammunition columns which supplied both infantry and artillery munitions were drawn from the establishment of artillery regiments. Field Artillery units provided divisional trains, and Foot Artillery the various corps and army columns.

Owing to the increased load placed upon the columns by conditions of warfare, the system was slightly altered in so far as all ammunition columns were placed under the command of the divisional train echelons, who drew ammunition from corps or army columns who in turn were supplied from railheads by the Army Intendents.

In April 1916, all vehicles were reduced to a four horse standard; prior to this, infantry columns were on a six horse basis.

At the beginning of 1917, the infantry and artillery columns were all broken up and were replaced by a new pattern of ammunition column. These, together with the light ammunition columns, were regrouped under the divisional train echelons.

LEFT: The forage wagon for the Leichte Feldhaubitze 98/09 was given higher sides to increase its carrying capacity. Subsequently this pattern forage wagon was also used by C 96 n/A and FK 16 batteries as a replacement for the type shown on the previous page. On this enlarged wagon the tailboard dropped to form a workbench and is seen here fitted with a vice.

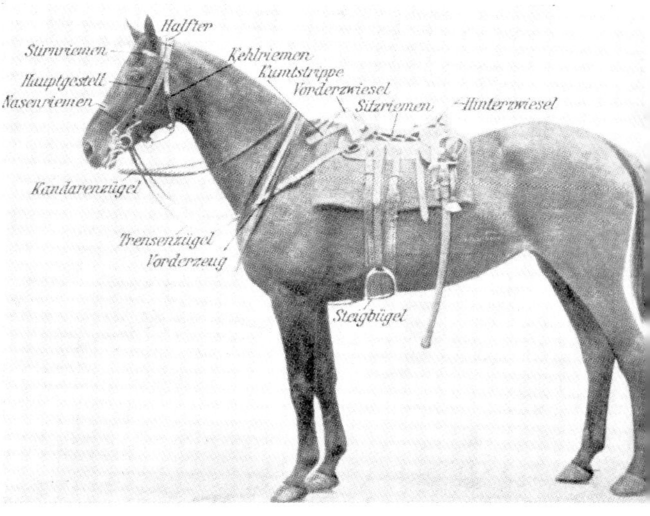

BELOW: A general view of the high-sided forage wagon as used by l.F.H. 98/09 and, later, C 96 n/A and FK 16 batteries.

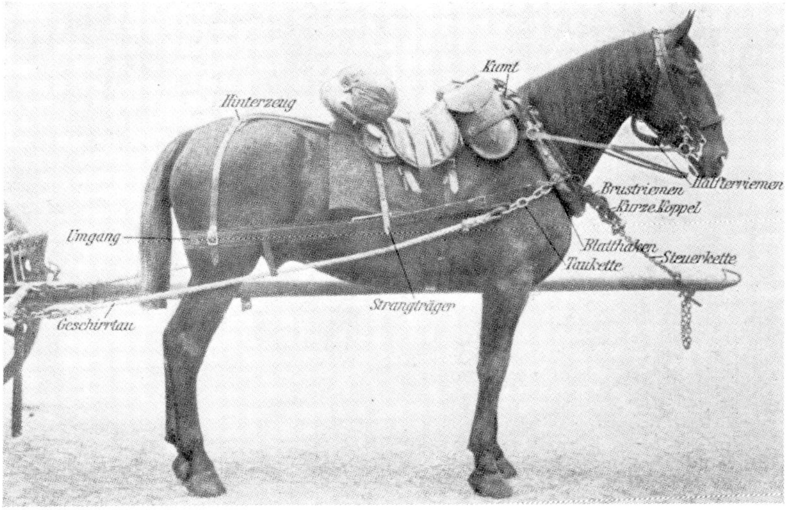

ABOVE: Harness and funiture for a lead horse of a gun or wagon. Main features as visible here were leather valises each side at the front of the saddle, and a rolled cover for the horse at the rear of the saddle. LEFT: Riding horse for artillery officer or senior NCO. BELOW: The stores wagon used by field artillery batteries. In some instances this type of wagon was used as a substitute for the standard forage wagon also. It carried spare wheels for the forage wagon and general battery stores.

WAR ESTABLISHMENTS OF AMMUNITION COLUMNS

Unit	Officers	Other Ranks	Horses	Vehicles	Armament: Rifles	Armament: Carbines	Armament: Revolvers	Ammunition Carried: Small-Arm	Ammunition Carried: Artillery: Shrapnel	Ammunition Carried: Artillery: High Explosive
LIGHT AMMUNITION COLUMN	4	188	178	24	—	--	180	—	1,080	3,810
HEAVY AMMUNITION COLUMNS										
ARTILLERY	3	180	187	26		68	22		2,061	468
INFANTRY 6-HORSED	3	181	191	26		66	23	662,400		
INFANTRY 4-HORSED	3	197	195	37		77	29	765,000		
ABTEILUNG of HEAVY AMMUNITION COLUMNS	26	1,129	1,158	181						
LIGHT AMMUNITION COLUMN of CAVALRY DIVISION	4	147	208	26			134	129,375	900	360
LIGHT AMMUNITION COLUMN of a BATTALION of HEAVY ARTILLERY of the FIELD ARMY	5	261	193	29	146		16			864
FOOT ARTILLERY AMMUNITION COLUMN	2	103	99	20	39		11	4,500		612
ABTEILUNG of AMMUNITION COLUMNS of a HEAVY FIELD HOWITZER BATTALION	23	839	810	163						
ABTEILUNG of AMMUNITION COLUMNS of a MORTAR BATTALION	25	427	414	83						
LINE of COMMUNICATION AMMUNITION COLUMN	3	133	136	53	55		13	1,575,000	or 5,400	

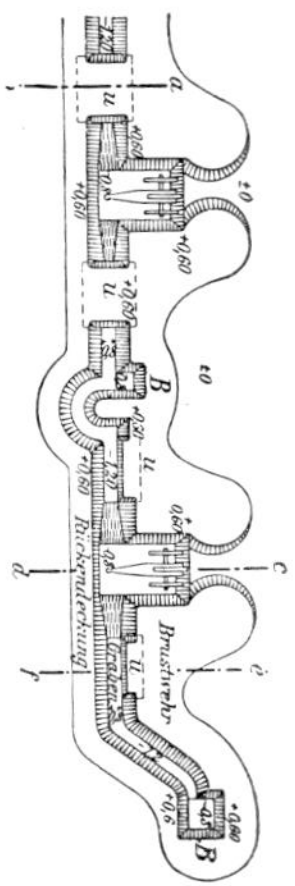

RIGHT: Layout for section of a field battery in a defensive line. B: observation posts (see page 11 for details): u: covered section of trench for ammunition storage.

CHAPTER 6
OTHER ARTILLERY UNITS

INFANTRY GUN BATTERIES

THESE units were a direct consequence of the advent of the tank, although it would be untrue to call them anti-tank artillery. Initially they were manned by a mixture of gunners and infantrymen and were equipped with an assortment of light field pieces of 3.7cm, 5.7cm (Belgian), and 6cm calibre.

During 1917, a series of 50 infantry gun batteries were formed and armed with captured Russian 7.62cm field guns. Shortly afterwards another 50 units, designated *Nahkampf-Batterien* were raised and were equipped with the 7.7cm field gun, mounted on low wheels. This weapon, the *7.7cm I.G. L/20* was soon replaced by the more effective *7.7cm I.G. L/27.*

Both types of battery consisted of 2 officers, 60-70 other ranks and 6 guns. They were usually allotted one to each divisional sector on active fronts. By the end of 1917, the 7.7cm infantry guns, considered unsuccessful in this role, were withdrawn and replaced by the light calibres mentioned above.

Numbers of these batteries were allotted to army assault battalions, being used in support of trench raids and other special enterprises.

MOUNTAIN ARTILLERY

NO mountain artillery units existed in the pre-war army, although material was available for their formation. About 25 batteries were formed during the war, being mostly grouped into 3 battery battalions, serving on the mountainous fronts in the Balkans, Carpathians, Alps and Vosges. The 8th, 16th, 17th and 21st batteries were armed with mountain howitzers and were attached to the 2nd (Bavarian), 5th, 6th and 7th Mountain Artillery Battalions respectively. The gunners were mainly drawn from Bavaria, Würtemberg and Baden.

A mountain battery, as laid down in the regulations of December 1916, contained four 7.5cm QF mountain guns, two in each of the battery's sections. The establishment of a section was 2 lieutenants, 1 sergeant major, 6 unteroffizier, 1 mounted orderly, 26 gunners, 31 drivers, 2 train drivers, 31 mules and 10 riding horses.

The principal weapon of the mountain artillery was a 7.5cm QF mountain gun. This had the recoil buffer placed beneath the barrel and was fitted with a single motion wedge breech block. The complete gun could be split down into seven loads, each being carried by one mule.

OBSERVATION AND SOUND-RANGING DETACHMENTS

THESE units were wartime formations, attached to armies, and were allotted as required to corps and divisions, where they were then placed under the command of the local artillery commander.

The observation detachments *(Artillerie-Mess-Truppen)* consisted of four or five flash spotting posts and a headquarters; the sound ranging units *(Shall-Mess-Truppen)* were organized on a similar basis, having as a basic unit a sound ranging post. The strength of both types of unit was 6 officers and about 100 other ranks.

These detachments worked in close co-operation with each other and the duties included the location and plotting on maps of hostile batteries' positions and the collation of all reconnaissance reports from their area.

The School of Instruction for observation personnel was at Wahn.

APPENDIX 1

EQUIPMENT FINISHES AND MARKINGS

THE traditional colour of Prussian artillery equipment was light blue. This rather conspicuous colour was changed in the early part of the 20th century to a pale blue-grey.

Prior to the war, in common with all wheeled equipment, guns had a unit designation stencilled on to them. These markings were to be found on the right side of the trail and were made in white paint. These unit designations took several forms but usually stated the regiment and the battery. A typical marking would have been: *FAR 21 2 Battr.* (Field Artillery Regiment No 21, 2nd Battery). Additional information was sometimes included. The Guard Foot Artillery Regiment, for instance, had the inscription *Garde Fs. AR* above a battery designation, *9 Battr (s FH),* indicating a heavy field howitzer. This same marking had the numeral 2 painted in the centre below both lines. This latter digit indicated the gun number within the battery, and was painted on to the back of the shield if the piece was fitted with one. The gun numbers were from 1 to 4, or 1 to 6 in field batteries. With the exception of gun numbers, the above markings were painted out upon mobilization, gun numbering being continued in many units throughout the war.

Although not common practice, personal inscriptions were used to a limited extent. They were painted on to the side of the barrel or on the muzzle end of the cradle. One example of this practice was the word *'Doria'* in large Gothic script painted in white on to the barrel of a *15cm K,* captured in August 1918.

The peace-time grey finish of artillery was soon replaced by field-grey. By 1917, many pieces were painted in camouflage patterns, a practice which was almost universal by the end of the war.

Many of these camouflaged pieces were painted in the field and there was very little conformity in style from gun to gun. Camouflage patterns fell into three basic types:

(1) 'Wavy' camouflage which broke up the shape of a gun by using different coloured fields of paint. Each field had wavy edges and its thickness was uneven.

(2) Broken camouflage, which was effected by painting an uneven

pattern of lines and curves across the entire gun in yellow or black and then filling in the resultant segments with different coloured paints.

(3) The third design was painted crudely on to the field grey factory finish and consisted of a two colour system of light and dark 'blotches' executed in rough lines across the gun's surface.

The colours used in camouflage painting were white, yellow, ochre, dull red, brown, green, grey and black.

APPENDIX 2

ARTILLERY UNIFORMS

THE field grey uniform was introduced as the service dress of the German Army in 1910. Essentially the uniform was of a camouflage nature but it incorporated all the traditional features of the old blue uniforms.

Service dress consisted of a field grey tunic, trousers and field cap; brown leather boots and equipment and the old full dress *pickelhaube* with a drab cloth cover.

The tunic was single-breasted, closed by 8 metal buttons. The cuffs were Swedish for field artillery regiments and Brandenburg for foot artillery units. The front edge of the tunic was piped in scarlet, and the collar and cuff in black, the traditional facing colour of the Prussian technical troops. In front of the skirts were pockets with flaps that slanted downwards from front to back, and at the rear were false pockets outlined in scarlet piping and decorated with buttons. The shoulder straps were detachable and were coloured field grey; the piping and cypher or number indicated the unit.

Trousers were field grey with a thin line of scarlet piping running down the outside seam.

The field cap was made of field grey material with a black cap band. Scarlet piping ran around the top and bottom edges of this band as well as around the crown rim seam. On the front of the cap were two circular metal cockades, on the crown the *Reichscocade* and one the band the State cockade.

Two items of the old full dress unifrom were worn on service. The light grey greatcoat and the leather *pickelhaube.*

The *pickelhaube* was surmounted by a ball for all regiments except those on the Bavarian establishment who had a spike. The helmet was black and highly polished. On the front was borne a metal plate which varied from state to state; on the sides were cockades mounted on the lugs that carried the chin straps, *Reichscocade* on the right, State cockade on the left. The whole was covered by a linen or canvas cover which prevented reflections, in the front of which in scarlet was the regimental number.

Although full equipment was issued to artillerymen they were seldom to be found wearing it. More normal was a belt with a bayonet, water-bottle and bread bag suspended from it.

LEFT: The distinctive helmet of the German Artillery, a pickelhaube, surmounted by a ball. Only Bavarian gunners wore a spiked helmet. The leather helmet is that of a Baden Field Artillery Regiment, the plate being the Baden Griffin. RIGHT: Second head-dress is a filzhelm, a wartime development of the pickelhaube made of compressed felt and of inferior quality. On active service a field grey or rush green cover was worn over leather helmets.

Changes in dress were more in the nature of trends within the entire army rather than specifically within any type of unit. Most changes were occasioned by economic necessity or war time practicality.

Pickelhauben underwent extensive changes during the period 1914-16, until their final replacement by the steel trench helmet. The number on the cover was changed to green and several ersatz types were used.

In late 1915 the entire dress of the army underwent a drastic 'paper' change. New patterns of clothing were authorised but these did not come into general use until 1917 in the artillery and even later in front line infantry units. In the transitional period from 1915-17 clothing which combined the new and old features was prevalent throughout the army.

The principal innovation of the new dress regulations was a field tunic called the *Bluse,* which had plain turn up cuffs, concealed buttons and a single square cut vent at the back. The shoulder strap piping and insignia were simplified, trade badges were not to be worn in the field, and all leather equipment was blackened.

The steel trench helmet was only issued to gunners after the requirements of infantry units had been fulfilled; this was in early 1917.

The greatcoat colour changed from grey to field grey, and in 1918 a new style of field cap was issued. This was of the old style with a patent leather peak and chin strap added.

Officers tended to wear clothing of a superior quality to other ranks, until ordered to wear the same in 1915. The field cap had a peak, and riding breeches were almost universally in use. Officers' greatcoats were dark grey, but later became the same shade of field grey as those of the other ranks. In 1914 swords were carried, but were soon dispensed with. They were hung from a belt woven in State colours. The insignia of rank was carried on the shoulder straps which were made of twisted cord sewn on to a stiff board.

Reserve units tended to wear obsolete equipment, being easily identified by the pre-1908 square leather cartridge boxes, and the long bayonet. The scarlet numeral on the helmet cover was surmounted by a roman R.

Certain units, guards for example, wore distinctive lace on the collars and cuffs, later reduced to the collar only on the *Bluse.*

Examples of all styles of artillery dress, and typical shoulder straps, are given in the colour plates.

PRINCIPAL WEAPONS USED BY THE GERMAN ARTILLERY, 1914-18

Official Abbreviation		Translation/description
	3·7cm Rev. K.	Revolver trench gun.
	3·7cm Grabenkanone	Trench gun.
	5cm K. (i. P.L.)	Trench gun in shielded mount.
russ.	**5.7cm K.**	Trench gun (Russian).
	7·5cm Geb. K.	Mountain gun.
russ.	**7.62cm I.G.**	Infantry gun (Russian).
	7·7cm I.G. L/20	Infantry gun.
	7.7cm I.G. L/27	Infantry gun.
	7·7cm I.G. L/19·5	Infantry gun.
	7·7cm I.G. 18	Infantry gun.
	7·7cm C. 96 n/A. or 7·7cm F.K. 96 n/A	Field gun.
	7·7cm F.K. 16	Field gun.
	9cm K. 73/88	Old field gun used by Foot Artillery.
franz.	95mm K.	Field gun (French)
	10cm K.	Heavy field gun.
	10cm K. 97	Heavy field gun.
	10cm K. 04	Heavy field gun.
	10cm K. 04/14	Heavy field gun.
	10·5cm I.F.H. 98/09	Light field howitzer.
	10·5cm F.H. 16	Light field howitzer.
	10.5cm I.F.H. Kp	Light field howitzer (Krupp).
	10·5cm S.K. L/35	Naval gun on wheeled carriage
	s. 12cm K.	Heavy gun.
franz.	lg. 120mm K.	Long gun (French).
	13cm K.	Heavy gun.
	15cm s.F.H.	Heavy field howitzer.
	15cm s.F.H. 02	Heavy field howitzer.
	15cm s.F.H. 13	Heavy field howitzer.
lg.	15cm s.F.H. 13	Heavy field howitzer (long).
	15cm K. L/30	Heavy gun.
	15cm S.K. L/40	Naval gun on wheeled carriage.
	15cm S.K. L/45	Naval gun on wheeled carriage.
	15cm Vers. K. Rh.	Experimental gun (Rheinmetal).
	15cm K. 16. Rh.	Heavy field gun (Rheinmetal).
	15cm Vers. i.R.L.	Experimental gun on wheeled carriage.
	15cm K. 16 Kp.	Heavy field gun (Krupp).
	15cm K.i.S.L.	Naval gun in Ships Mounting.
	15cm R.K.	Ring cannon.
lg.	15cm R.K.	Long ring cannon.
	17cm S.K. L/40	Naval gun.
	17cm K.i.R.L.	Naval gun on wheeled carriage.
	17cm K. Bett-Gesch.	Emplaced Naval gun.
	17cm K. Eis	Railway gun.
	21cm Mrs.	Heavy howitzer (1902).

(continued)

Official Abbreviation	Translation/description
Mrs.	Heavy howitzer (1910).
21cm lg. Mrs.	Heavy howitzer (1910).
21cm S.K. L/40	Heavy emplaced Naval gun.
21cm S.K. L/45	Heavy emplaced Naval gun.
21cm K. Eis	Railway gun.
24cm S.K. L/30	Emplaced Naval gun.
24cm S.K. L/40 Eis	Railway gun.
24cm S.K. Bettgesch	Emplaced Naval gun.
28cm Hbtz. L/12	Heavy howitzer.
28cm Hbtz. L/14	Heavy howitzer.
28cm Kst. H. L/16 i Kst. H.L.	Heavy coastal defence howitzer.
28cm S.K. L/40	Emplaced Naval gun.
28·5cm S.K. L/45 i Kst. P.L.	Heavy Coastal defence howitzer.
28·5cm S.K. L/50 i Kst. P.L.	Heavy Coastal defence howitzer.
30·5cm S.K. L/50	Heavy Naval gun.
30·5cm Mörser Kp. 98	Heavy howitzer (Krupp).
30·5cm Mörser Kp. 09	Heavy howitzer (Krupp).
30·5cm Mörser	Austrian heavy howitzer 1911 model
30·5cm s. Kst. Mrs. 96	Heavy coastal defence howitzer.
30·5cm s. Kst. Mrs. 09	Heavy coastal defence howitzer.
35·6cm S.K. L/52	Heavy Naval gun.
35/38cm S.K. L/45	Heavy Naval gun.
38cm S.K. L/45	Heavy Naval gun.
42cm Mrs.	Heavy howitzer.
42cm Kz. Mar. K.	Short Naval gun (howitzer).
42cm Kz. Mar. K. 14.	Short Naval gun (howitzer) on wheeled carriage.

NB: The designation L/40 etc, etc, indicates the length of the barrel expressed in Calibres (ie, 40 x calibre = length)